Basic Medieval and Modern
Military Tactics for Authors

HOW TO WRITE A FANTASY BATTLE

SUZANNAH ROWNTREE

Cover design by MiblArt.

To the memory of two women:
Christine de Pizan, the first woman to make a living by
her pen, who in 1410 published a treatise on arms and
fortifications;
and Marie von Clausewitz, who worked as a researcher
and editor on her husband's landmark book On War,
and who brought the project to completion and publica-
tion after his death.

Introduction

S o you want to write a fantasy battle. And you're feeling a little bit intimidated.

You don't know a lot about military tactics. You couldn't tell a mangonel from a trebuchet even if your fief depended on it. You're no good at chess. You're not the kind of person who cares deeply about tanks. And, you'd try to learn something, but either you just don't get on well with non-fiction, or you haven't a clue where to start. The internet is full of neckbeards who will laugh you to scorn for your uphill heavy cavalry charge, but, fingers crossed, they're never going to read your book. You asked your cover artist to put a princess on the front, with a sword and a pretty dress, specifically to warn them off.

All the same, you've got this big battle scene coming up at the end of the book, where the princess, the dark wizard, and the trusted allies are finally going to team up to defeat the *real* bad guys. It's going to be epic. It's going to be magical. It's going to be...what, exactly?

You have no idea what needs to happen. How does a battle get fought?

Yep. We've all been there. I assure you that none of us emerge from the womb with an instinctive knowledge of battle tactics. We all have to start somewhere, and that's why I've written this little book.

My goal is to be short, easy to read, and give you just enough information to tackle that battle scene with confidence, and with an idea of where to go looking for more if you need it. First, though, I want to reassure you that if I can understand these things, then you can, too.

I knew basically nothing about military tactics ten years ago, when I decided that I wanted to write a historical fantasy based on the two hundred year history of the medieval crusader states. I wasn't even particularly interested in military tactics at the time, but it turns out that if you spend the best part of a decade obsessively reading and trying to understand medieval history, especially the bits with wars in...well, you end up picking up quite a lot of information on military matters. And then you end up liking it for its own sake.

And then, one fine February day, a dictator in another hemisphere decides it'd be cool to invade a neighbouring country. As the largest-scale land war of your lifetime unfolds amidst unspeakable horror and destruction, you begin to realise just how *much* of what you know about medieval warfare is still true, to some extent, today.

Not everyone has the same experience as I have. Many people undoubtedly have more. In the interests of full disclosure, I have neither a history degree nor any formal military training. My expertise, such as it is, is closely focused on warfare during the high middle ages, approximately AD 1000-1300, and I'm weak on post-industrial warfare. Since this is a short book, it will also contain a number of wild generalisations, to say nothing of gaps—I've spent very little time discussing either siege or urban warfare, for instance.

(Another area that I will not be discussing in this book is that of wound care, field hospitals, and camp hygiene in medieval camps and on medieval battlefields. For information on these topics, please consult Piers D. Mitchell's book *Medicine in the Crusades: Warfare,*

Wounds, and the Medieval Surgeon, which is thoroughly readable and engaging.)

All this is to say that by far the best way to learn anything about this topic is not to stop with this book, but to go on and read some resources from people who actually know what they're talking about.

The two books on medieval military history that have taught me the most are John France's *Western Warfare in the Age of the Crusades, 1000–1300,* and *Victory in the East: A Military History of the First Crusade.* Both of those books are available for loan on the Internet Archive. I can also suggest *The Origins of War: From the Stone Age to Alexander the Great* by Arther Ferrill if you're working with a more ancient, Bronze-Age style of warfare.

There are also classic works on military tactics which have influenced real battles throughout history. *De Rei Militari* is a Roman treatise, by Vegetius, which was highly influential through late antiquity and the Middle Ages. A more recent work is *On War* by Carl von Clausewitz, first published in 1832 and still extremely influential today. From the twentieth century, two standard works are *Strategy* by BH Liddell Hart and *Military Misfortunes* by Cohen and Gooch. I have to admit that to date I have read only the last of these. (And it was fascinating!)

Most of what I've included in this book is based on my study of individual battles, mostly from the crusades. A familiarity with real battles is going to be invaluable, and the reason is simple: every battle is absolutely unique. Each and every battle that has ever been, has been fought for different reasons, on different terrain, with different weapons, under different weather conditions, with different commanders making different gambles and different mistakes. It is by studying individual battles that you'll truly begin to comprehend the diversity of what can happen in war. For this purpose, I would strongly recommend familiarising yourself with the Osprey Campaign series. This series contains 391 works by military historians, and while I haven't read a lot of them, every single one I *have* read has been invaluable. They are short books, full of detailed illustrations and pretty diagrams, providing scholarly yet readable accounts of battles from all across history. If you're writing a fantasy battle with, say, a medieval-style setting, you could do much worse than look up a few of the Osprey Campaign books about medieval battles and familiarise yourself with the contents.

That said, while all battles are unique, there are also some basic principles that tend to remain the same across time and space—and will also apply to your fantasy battles, since your magical elements will naturally correspond to some element of real-world warfare. This little book is intended to tell you about these points of commonality.

Let's begin—but not with the battle itself. First, we'll need to back up a bit and lay some foundations.

Societies at War

Not all societies and cultures are the same as each other, and certainly they don't all resemble modern Australian (or UK, or American) culture. Their beliefs, technology, resources, and economies may all be radically different, and all of these things will have an impact on how they make war. As you plan your fantasy, you'll need to spend some time thinking about the different ways that your fantasy culture's world-building will matter for your battle scenes.

For instance, here are some examples from real history.

During the Middle Ages, the way the Italian merchant cities like Venice, Pisa, and Genoa made war was quite different to the way that the feudal lords of northwestern Europe made war. In northwestern Europe—specifically France, England, and Germany—the economy was mostly land-based. Agriculture was the source of wealth. Technology moved very slowly, making it impractical to increase the land's yield in that way. As a result, if you wanted to get richer, you had to expand the amount of land whose resources you could exploit. As a result, petty warfare between landowners became endemic. People were constantly carrying out small-scale raids to weaken their neighbours' control of their land, while building castles to consolidate their own control of their own land.

In the Italian trading cities, however, the source of wealth was quite different. Access to the Mediterranean Sea meant merchants could travel far and wide, buying African gold, Syrian glass, and Byzantine silk—then bring it to the Italian ports from whence it would make the long journey through the mountains to the hungry markets of northwestern Europe. Foodstuffs, slaves, and livestock could then be brought back to the trading ports and set out on their way to markets around the Mediterranean. Expanding your wealth was therefore about expanding your trade, not your land. As a result, rather than owning tracts of land dotted with formidable castles, Italian nobles lived together in luxury in their own cities in unfortified villas. Economic realities made it profitable for them to work together with their neighbours rather than to be constantly squabbling. When they did go to war, it was usually a naval war rather than a land war. It was usually a war with one of the other big trading cities—like the thirteenth-century War of St Sabas, which pitted Venice and Pisa against Genoa and was fought all around the Mediterranean for decades. And the prize was not land—it was trading rights and quarters in the big trading cities: Alexandria, Acre, Constantinople, and others.

Since war was of primary importance to northwestern Europe, being the main way of increasing wealth, the lords of northwestern Europe were all trained as warriors from an early age. But since the main way to increase wealth in Italy was trade, a noble's primary occupation would be as a merchant, not a warrior. So, in Italy warriors became specialised and monetised. As the middle ages gave way to the Renaissance, land wars in Italy were fought by mercenaries, or *condottieres*. Whole armies would hire themselves out to whoever could afford to employ them, often changing sides whenever the money ran out. Two cities could go to war and never put any of their own men on the battlefield. Not only that, but the mercenaries had little incentive to die or be killed for their employers. What they *were* motivated to do was to earn money—and one of the most profitable ways to fight a battle was to capture as many of the opposing army's warriors as you could, then hold them to ransom. Some battles were almost entirely bloodless, determined by how many prisoners could be captured by either side.

Another society that fought quite differently to the feudal lords of northwestern Europe was the Byzantine Empire. Whenever these two cultures faced off on the battlefield, the French or Norman lords would shake their heads in disbelief. Their idea of honour was tied to warfare: you proved yourself respectable by fighting, bravely and often. This was not as important to the more pragmatic Byzantines. While everybody avoided pitched battle if they could help it, even the feudal knights of western Europe, the Byzantines considered it perfectly honourable to do so through deception and cunning. Whenever the Byzantine Empire was invaded, even at the height of its power, the emperor's first instinct was often not to respond with armed force but with negotiators. Invading armies need to be supplied by friendly populations with food, weapons, animals, and medicines. In enemy territory, far from such friendly populations, their health and strength is precarious. So in the best-case scenario an invasion could be very cheaply and bloodlessly defeated just by spinning out negotiations until domestic troubles, an epidemic, or the onset of winter sapped the invaders' strength. That's when the Byzantine army would lumber into motion, easily defeating their weakened opponent now that they had thoroughly lost the initiative. To the European knights, such behaviour was devious and cowardly. But it usually worked.

As world history marched on, technological innovations began to radically change the ways wars were fought. In the early modern period, Europeans developed firearms, artillery, and highly-trained professional armies. Once these new weapons and tactics had been honed in warfare against similarly developed opponents in Europe, a window of opportunity opened in which modern European warfare had a crushing advantage compared to armies all around the world—in India, Africa, and China. Without this military advantage, the age of imperialism probably wouldn't have got off to such a strong start. If stark technological, tactical, or magical power imbalances play a role in your fantasy worldbuilding, then this is the time period you probably need to study. Try reading William Dalrymple's *The*

Anarchy: The East India Company, Corporate Violence, and the Pillage of an Empire for an introduction to military history in this period.

Ethical and political considerations can also make a huge difference to the way a society fights. We've seen these as linked factors in the present Russo-Ukrainian war. Russia is a country with a strong autocratic tradition, which has never yet created a strong or healthy democracy. From the tsarist regime to the Communist regime to Vladimir Putin's thinly veiled dictatorship, the average Russian has usually been the helpless subject of a totalitarian ruler. As a result, the way the Russian army fights has traditionally been horrifically wasteful in terms of human life, relying upon frontal charges to overwhelm the enemy with the sheer weight of their numbers. The tsarist army fought this way in World War I, the Red Army fought this way in World War II, and that's the way Putin's army is fighting in Ukraine today, taking hundreds of thousands of casualties along the way despite rapidly declining Russian demographics. By contrast, in the thirty-odd years since the dissolution of the Soviet Union, Ukraine has developed a troubled but viable democracy. Its people are citizens, not subjects. Just this century they've been through two pro-democratic revolutions. Such people have agency;

they cannot be fed into a meatgrinder without serious political consequences. Accordingly, Ukrainian generals have adopted vastly different tactics when confronting the enemy. Instead of frontal meatgrinder assaults, the Ukrainian Armed Forces focus on precision distance strikes using advanced drone systems and artillery. Their command structure is similarly democratised, a crucial feature that allowed them to survive the first, desperate days of the 2022 Russian invasion.

Russia's army still fights using an old, centralised Soviet command style. "Key decisions are made at the very top, even local operations require layers of approvals, and initiative is often punished rather than rewarded," writes Yaroslav Trofimov in his history of the first year of the invasion, *Our Enemies Will Vanish: The Russian Invasion and Ukraine's War of Independence.* Russian infantry who penetrate Ukrainian lines, for instance, must go to ground pending fresh orders from higher up. In the autocratic, hierarchical Russian command structure, there's a great deal of pressure not to annoy the higher ups by reporting bad news. This results in an army culture in which the commanders are ill-informed and the whole army is slow to adapt and respond to conditions on the ground.

Ukraine, by comparison, had recently shifted its command style from this Soviet style towards a more decentralised, NATO-standard "mission control" style. "Once the top commander's overall intent was clear," Trofimov explains, "units were free to execute their missions as they saw fit, without asking for additional permission." Such a command style relies upon commanders who trust their subordinates and subordinates who are capable of taking initiative and responsibility—distinctively democratic qualities. It fosters a more agile army, capable of acting autonomously in the face of multiple threats. In the first critical days of the invasion, with the enemy advancing on ten different axes, a Soviet top-down command style would have broken down under the sheer pressure of having to personally handle so many vital decisions.

So, as you start planning your fantasy battles, it's probably wise to begin by thinking about your combatants' fighting style. How have they been shaped by their economies, their sense of honour, their technological or magical resources, and their ethical or political considerations? All these things will have an impact on how they fight.

Tactics or Strategy?

The *Oxford English Dictionary* defines "tactic" as "the art of disposing armed forces in order of battle and of organizing operations, especially during contact with an enemy." So, the use of various battle formations like the Ancient Greek phalanx, the Roman tortoise, or the medieval massed cavalry charge with couched lances are examples of tactics.

According to the same dictionary, "strategy" is "the art of planning and directing overall military operations and movements in a war or battle." So, strategy is a higher level of planning—for example, that Byzantine trick to win the invasion by distracting the enemy until an epidemic makes it impossible for them to fight.

To massively oversimplify things, tactics are what you use to win a battle, and strategies are what you use to win a war.

Before the Battle

"Fail to plan, and you'll plan to fail" applies in the military realm just as it does everywhere else. In real life, battles usually don't happen by mistake. In fact, I *am* aware of one battle—the Foraging Battle of 31 December, 1097 during the First Crusade—in which a small force of Crusaders looking for food accidentally bumped into a Turkish army on its way to confront them. The result was a victory for the Crusaders—but only barely, and largely due to the fact that they had with them their most brilliant commander, Bohemond of Taranto, who was able to come up with a plan of battle on the fly. But this was a stroke of luck for them, and the victory cost them dearly.

Battle isn't a pub brawl: the stakes are not just one or two lives, but hundreds. The higher the stakes—a city, a kingdom—the more carefully your fantasy leaders will want to plan the battle. This planning phase is actually so important that it can itself become a determinative factor in the battle. We all know that whoever can seize the element of surprise usually gains a huge advantage. And surprise is powerful precisely because it denies the enemy that crucial time to plan. In *Military Misfortunes,* Cohen and Gooch identify three basic mistakes that can be made in war, and one of them—Failure to Anticipate—is largely about falling into the trap of surprise.

Add to this the fact that in pre-industrial war—war before technological advances became common and determinative—the outcome of a pitched field battle was almost impossible to predict. Before the invention of firearms, factors like numbers, training, and equipment could have an effect but were rarely determinative on their own. Rather, John France concludes in his book *Western Warfare in the Age of the Crusades,* the most determinative factor was an intangible one: which side had the strongest expectation of victory? It was this side that was most likely to succeed on the medieval battlefield.

To be honest, intangible factors are still super important in modern warfare. In the first days of Russia's invasion of Ukraine, everyone I knew in Australia and America was betting on Ukraine falling apart within weeks if not days. But if you listened to the Ukrainians themselves, they seemed fully confident of fighting and beating Russia. For me, this was significant because of what I knew about medieval warfare. Indeed, you have only to look at the past century or two in the history of Vietnam or Afghanistan, to see undersupplied yet motivated forces overcoming more technologically advanced empires time and again.

Pitched Battle, Siege, Skirmish, or Ambush?

A pitched battle is a battle fought via manoeuvre in the open field, as opposed to one fought in or around stationary fortifications—which is usually known as a siege. A pitched field battle makes use of terrain and manoeuvre, as both sides seek to position themselves on the landscape in such a way as to give themselves an advantage. A pitched battle also usually refers to a large-scale engagement between two massed armies for high stakes, as opposed to a skirmish or raid, which are carried out by smaller forces to disrupt enemy warriors or farmers, and usually not intended to make lasting territorial gains. Finally, a pitched battle is also usually a battle of choice, in which each side has the option to back out but chooses not to. This can be distinguished from an ambush, which is forced by one side on the other at either a place or a time not of their choosing.

In 21st century warfare, you're likely to see commentators referring to positional warfare or manoeuvre warfare. Positional warfare, like an old-fashioned siege, is characterised by relatively static front lines and small, incremental gains—think of classic WWI-style trench warfare or the more recent Battle of Bakhmut. Manoeuvre warfare is somewhat analogous to old-fashioned field warfare, with both sides able to move more freely about the theatre of war—for example, World War II's Battle of Kursk or the Ukrainian Kharkhiv offensive of September 2022. While positional warfare is uniquely unexciting and unpleasant, it's still an important element in modern warfare that can be used to consolidate gains made

> during a manoeuvre phase, to wear down the enemy, or to stabilise a battle front preparatory to going on the offensive.

Since it was almost impossible to predict victory, the average medieval commander usually saw pitched battle in the field as something to be avoided at all costs. Even if you didn't resort to Byzantine-style strategy, delaying a confrontation until your opponent's health and supply lines broke down, it was still far safer to avoid a large-scale confrontation.

As a result, if you wanted to take your opponent's territory, you might gradually erode his (or her) authority and power by continually raiding his lands—carrying away prisoners, food, and livestock. Do this diligently, and the territory ceases to be profitable to your opponent. Incomes fall, defence costs go up, and his vassals start to think about switching sides just to get some peace. As his grip weakens, you start to build castles on his land. Now, you have a strong place for your warriors to hide if he tries to evict them, which also doubles as a beachhead from which you can control the local peasants while launching more raids further into your enemy's territory. Bit by bit, you'll gain your objective, without ever having to risk a pitched battle. This is additionally sensible, because although your vassals—your underlords—have made solemn vows to come to your aid with warriors whenever you need to fight, realistically in the decentralised medieval feudal system they are not fully under your control. You never know if you can really rely on them. What if you threw a battle and none of your own men came? Much safer to rely on small-scale raiding.

But, let's say that a pitched battle has become necessary. It's not just a neighbouring lord who wants to take a bit of your land—it's a neighbouring empire, and their army is massive, and it's time for your people to stop raiding each other and pull together to defeat them.

You'll still want to avoid giving battle. Can you negotiate a settlement? Sometimes you can't. Some of your lords may feel that this would be dishonourable, or perhaps the enemy has genocidal intentions which make a negotiated settlement impossible. Or perhaps the enemy simply has no intention of honouring its agreement. At the dawn of the eleventh century, Ethelred the Unready famously found himself in this position. Ethelred negotiated the payment of "Danegeld"—a large sum of money—to the Danish pirates who regularly plundered England, only to find himself handing over ever larger sums on a regular basis. As Rudyard Kipling quipped, *once you have paid him the Danegeld, you never get rid of the Dane.*

Can you defeat your opponent by buying time, like the Byzantines, or decoying her into a trap? Maybe, if you let her advance far enough into your territory, you can move in behind her and cut off her supply lines, forcing her to surrender. This is what happened to the Seventh Crusade—the crusaders were trapped in Egypt when the Nile flooded, cutting off not only their supplies but also their retreat. The entire expedition had to make an ignominious surrender.

Or maybe you can take up your position at a place where the terrain favours you, and your opponent will decide to go away again rather than fight you at a disadvantage. This is what Guy of Lusignan did to defeat Saladin in the 1183 Bethsan campaign—but failed to repeat four years later at the Battle of Hattin. At any point you might take refuge in some city or castle, bar the gates, and let the enemy besiege you. While he's freezing his toes off in the camp, you're waiting for your reinforcements or allies to turn up and swing the balance so overwhelmingly in your favour that with any luck, the enemy runs away rather than fight.

Sometimes even these tactics don't work. The risk of waiting for the enemy to make a mistake is that this cedes the initiative to the enemy: while you're delaying or retreating, he's moving forwards, taking plunder, destroying your people, gaining confidence and resources. Fighting defensively is safer and may buy you a vital advantage, but it doesn't win a war. If there are no reinforcements or allies to come to your aid, there's no point in letting yourself bleed out in a series of hopeless sieges. In this case it would be much better to accept the risk of pitched battle, lay your plans, pick your ground, and do your best to smash your enemy while you can.

Sometimes, too, people also choose to fight simply because they've vastly overestimated their own strength. They've underestimated the enemy's numbers. Or they thought they'd have enough supplies, and they didn't. Or they thought they'd have help from an ally, who in reality is just biding their time to disappear at the moment they're most needed.

As you're writing your book, don't forget to let your characters lay the groundwork for the battle. Especially in medieval or ancient style warfare, pitched battle is going to be their last resort, so we might see them debating among themselves whether seeking battle or avoiding it will give them the best chance of success.

Along with these discussions will also be various plans of battle—is it possible to stage an ambush? Which terrain will give us the best advantage? How can we deploy our forces to give us the best possibility of victory? Once battle has been decided upon, there may be further preparations to make—food and horses and allies to find, religious ceremonies to observe, forces to mass and deploy.

In all this, don't forget that your characters will also need to lay plans for what happens if the battle goes *badly* for them. If they're forced to retreat, to which rallying point will they fall back? Which reserves and what fortifications or features of the terrain will protect them when they do? And, if there's time to think that far ahead, what will they try next if they fail? Of course, given that you're writing a high-stakes fantasy novel, the big battle plan might not admit any failure. In *The Lord of the Rings* when the Riders of Rohan go to war in Gondor, they do so leaving their own land wide open to the attacks of the Enemy across the Anduin in their rear. Tolkien uses this for dramatic effect, showing his readers that this is an all-or-nothing battle: if the ride of the Rohirrim results in anything less than a complete victory, they will be toast.

On the other hand, perhaps your characters are hoping to avoid battle, but they're aware that the enemy is trying to force a confrontation on *them*. They might not have a specific plan for a specific battle, but they should always be thinking about what they are going to do in the event of a surprise attack. Does their road lead through a place that would be ideal for ambush? How defensible is this campsite, or would they be better off finding a different place to spend the night? How will they leave this castle if the enemy swoops down to surround it? What are the risks, and what things should be done to mitigate those risks?

Battle Objectives

As your characters plan their battle, don't forget to consider what their objectives are, not just on the strategic level, but also on the tactical.

Different strategic objectives will dictate different tactics. We usually think of warfare as having territorial objectives, but there are other reasons why people go to war, all of which affect the way they fight. Fighting to gain or secure territory will look quite different to fighting to survive, to kill the enemy, to drum up political support, or to win back prestige. In the 1973 Yom Kippur War, Egypt attacked Israel primarily to win back a sense of honour after their humiliating defeat in the Six-Day War. They lost the Yom Kippur War in a territorial sense, but humiliated Israel so badly with their surprise attack that they gained their real strategic objective, which was their sense of military pride.

On the other hand, it's all very well to have an overriding goal, like "let's not be conquered by the Dark Lord" or "let's not be slaughtered by genocidal maniacs" or "let's not allow our princess to be carried off by the unwanted suitor." But what strategies and tactics will your characters use to achieve whatever goal they have? Similarly, what tactical goals will their enemy be trying to achieve?

The Lord of the Rings contains some of the most accurate fictional depictions of medieval battle ever put to paper (for this, thank the fact that JRR Tolkien was both a medieval scholar and had lived through two world wars). Consider Aragorn's very different objectives at the Battle of Helm's Deep, the Battle of the Pelennor Fields, and the Battle of the Black Gates. At Helm's Deep, we have a classic siege. Unlike in the Peter Jackson films, which do not demonstrate a fraction of Tolkien's military knowledge, Tolkien has Aragorn and Theoden lead their forces to Helm's Deep to reinforce the garrison ahead of an expected siege. They arrive just in time and are besieged along with the small original garrison, which would have had no hope of holding out without them.

A Word About Sieges

OK, I can't hold myself back any longer. Television and movies hate sieges. Producers can't stand the thought of letting the heroes hide out in a nice, cosy, unassailable fortress while the enemy dies by inches outside. *"If we let our characters do the sensible thing, how will we fill our Swords Go Clang quota for this episode?"* So, the characters always come up with a ridiculously convoluted reason for why they shouldn't hide out in the fortress, which has been custom designed and built at vast cost precisely to protect them in this very situation. Instead, they wait around outside and let the enemy kill most of them before gaining a hard-fought victory they could have won far more easily if they'd just stayed inside and locked the door.

Ignoring fortifications is probably the number one most common mistake writers make writing their battle scenes, and it always makes me want to scream. Not only is it not true that sieges are lacking in drama or even Swords Going Clang (as Tolkien demonstrated pretty convincingly in the Battle of Helm's Deep), but the only reason you'd ever march away from your fortifications in order to fight would be if it was your only choice—if, for instance, you *knew* your reinforcements were not coming and your one last hope of survival would be to seize the initiative and ambush the enemy while they're still on their way to get you.

An additional problem is that even in their darkest hour, the TV and movie characters never think of converting their hopeless siege into a surprise attack. They just stand around aimlessly outside the walls twiddling their thumbs, waiting for the enemy to pile into them and slaughter them all. That's the second most common mistake, but we'll get to that later when we talk about surprise and terrain.

In choosing to weather Saruman's assault within the safety of Helm's Deep, Theoden and Aragorn are making by far the most sensible decision available to them. Saruman's army is strong, fresh, numerous, and has the initiative. Theoden's men have not yet been gathered, nor the country put on a war footing. He's in no shape to go head-on with a strong and concentrated enemy. Behind the walls of Helm's Deep, Theoden can keep his existing forces alive and safe while Gandalf rides to gather reinforcements from other parts of the kingdom. Meanwhile, the king's presence at Helm's Deep forces the enemy into a classic military dilemma. The enemy can't simply ride past Helm's Deep, leaving Theoden safely hidden inside—even though this would give them access to the capital, Meduseld, together with all the plunder and territory available within the kingdom—because that would give Theoden the opportunity to emerge from Helm's Deep and cut the enemy off from home base at Isengard.

(As a side note, something like this actually did happen at the 1177 Battle of Montgisard, when Saladin made the mistake of breezing right past the fortress of Gaza where King Baldwin IV and his small force of knights were hiding. King Baldwin waited for Saladin to go past, then promptly emerged from Gaza, launched a surprise attack from the rear, and dealt Saladin the most crushing defeat of his career. See? Tolkien knows exactly what he's writing here!)

On the other hand, if Saruman's army turns aside to besiege Helm's Deep, the defensive key to the northern part of the kingdom of Rohan, then they risk getting bogged down in a long siege if the defenders prove to be stubborn. They'll lose the initiative, bleed out in the attritive fighting, and become sitting ducks for when Gandalf and his reinforcements do arrive.

In fact, this is more or less what happens at Helm's Deep. Theoden's presence rallies the defenders to put up a spirited resistance. While intense fighting continues all night, Gandalf gathers the scattered troops defeated at the border—the Fords of Isen—where the enemy army entered the kingdom. He then marches them to Helm's Deep just as the sun is rising and the enemy assault on Helm's Deep has been repelled by the defenders. In the sunlight, the orcs of Isengard are depleted and quickly demoralised by the shock arrival of reinforcements. As they flee, the trees of Fangorn Forest turn up and ensure that there are no survivors. So, the tactical objective for Aragorn and Theoden is simple: maintain the defences of Helm's Deep, absorb the enemy's first attack and then hold on until reinforcements arrive to turn the tide. It's a *very* classic siege situation, and although it unfolds a little more quickly than would be normal in real life, it's also thoroughly plausible (give or take a few orc-eating trees).

The situation at the Battle of the Pelennor Fields is quite different, even though this battle also has to do with a siege. When the Dark Lord besieges the city of Gondor, its people have very little hope of reinforcement, and their objective is more or less only to maintain the defences for as long as possible before the city falls. But, thanks to the efforts of Gandalf and Aragorn at Helm's Deep, the neighbouring kingdom of Rohan is able to come to the aid of their allies. Aragorn then manages to play a trump card by taking a dangerous short cut south via the Paths of the Dead, where he captures the enemy's navy and brings reinforcements from down the river. By the time Aragorn arrives on the battlefield, the Rohir reinforcements have already arrived and made a surprise dawn attack on the massed besiegers. However, as devastating as the Rohir charge is, the enemy army is large enough and confident enough to absorb it without breaking. It's only once Aragorn arrives with his ships to launch a second, surprise attack from a new direction that the enemy army panics and begins to flee. Aragorn's objective, therefore, is not so much to outfight or outnumber the enemy (which would be impossible), as to break their will. Two surprise attacks from unexpected reinforcements, as well as the death in single combat of the enemy commander, accomplish this task.

Finally, at the Battle of the Black Gate, we have quite a different situation—a situation in which tactical sense is sacrificed to a larger strategic purpose. Aragorn and his massed allies march along Mordor's western border to its gate, blowing trumpets and playing the part of the big fantasy heroes who are about to save the world. He then lines up outside the Black Gate and challenges Sauron to fight. Through all this he uses very little finesse and absolutely no element of surprise. If he actually meant to fight and win a battle against Sauron, Aragorn would never have behaved in such an arrogant way—unless, of course, he'd vastly overestimated his own strength. Throughout *The Return of the King,* Aragorn behaves in a way calculated to convince Sauron that he wields the powerful One Ring. In fact, Aragorn does not have the Ring: he's playing a totally different game to the one Sauron thinks he is playing. On a strategic level, Aragorn is simply trying to distract Sauron from the real threat—two tiny hobbits, sick and starving and on foot, creeping with painful slowness into the heart of his kingdom to destroy the Ring. Aragorn's entire expedition, and the battle that follows, is a diversion calculated to draw Sauron's armies north, out of Frodo and Sam's way. He uses intentionally bad tactics in pursuit of a larger strategy, and Sauron falls for it, never realising that Aragorn's objective in *this* battle was never military at all.

Unfortunately, the film and TV crews adapting Tolkien have never measured up to his in-depth military knowledge. For instance, *The Rings of Power* is one of my favourite television shows ever, but the military tactics are pretty egregious. In season two, the Orcs under their leader Adar are engaged in besieging the Elvish city of Eregion when a relief army arrives, commanded by Elrond and the High King Gil-Galad. After initial negotiations break down, Elrond returns to the attack—only, instead of attacking Adar himself and his camp, he instead attacks the besiegers before the walls of Eregion.

The choice to ride directly onto the battlefield makes sense at the Battle of the Pelennor Fields, where Tolkien has Theoden and the Rohirrim move about the battlefield strategically targeting the enemy's camps and leaders. But in *The Rings of Power,* the main characters' actions don't make sense, as Adar's camp is located at a greater distance from the siege. At first, when Elrond arrives in the area, Adar seems to have enough reserve forces to block Elrond's path to the city, which his main forces are busy besieging. Shortly afterwards, when negotiations break down, Adar fails to use these reserves to protect the besiegers from Elrond, who in turn faces relatively little resistance as he leads his own army onto the battlefield around the walls of Eregion. Where did all Adar's reserves disappear to? Are they just sitting around twiddling their thumbs?

Let's presume that the reserves are defending Adar's camp. Elrond thus has two options here: either he could attack the camp, which contains Adar and the Orcs' non-combatants, including families. Or, he could attack the Orcs besieging Eregion and hope to drive them away from the walls of the city. He chooses the latter, but honestly I think this is the wrong move. By attacking the besiegers, Elrond risks getting bogged down in the fighting around the walls, losing the initiative, and ultimately getting caught between the besiegers and the fresh reserves—which does in fact happen. If Elrond used his wits, he would attack Adar's camp instead. He would have the chance to deal a decisive blow against Adar's freshest troops while his army is still at full strength. By threatening their leader, he would force the besiegers to retreat from the walls of Eregion to help defend the camp, thereby relieving the besieged city. And above all, he might kill or capture Adar himself, thereby depriving the enemy of their commander and

father figure and therefore the will to fight. But Elrond
fails to do this.

I could go on a lot longer about the military sins of
The Rings of Power (for one thing, a massed medieval
style cavalry charge is not something you can stop with
a single cry of "Halt"), but I'm going to spare you.
The point here is that a sensible commander will always
try to kill two birds with one stone. Just as in a game
of chess you can head off the attack of an opponent's
knight or bishop by threatening her king, so in warfare,
your best tactical objective will often be something *different* to your opponent's tactical objective. Distract
her; divert her; funnel her into a course of action or a
field of battle that *you* control.

We see an example of this in the present phase of the
Russo-Ukrainian war. Broadly speaking, the Russian
goal in the war has always been Ukraine's submission
as a vassal state of Moscow. In 2022 Russia attempted
a blitzkrieg to seize territory and topple the democratic Ukrainian government, but this failed within a
month and the Russians were forced to withdraw from
the capital, Kyiv. Russian goals then narrowed to focus on conquering more territory. Ukraine responded at first by seeking to liberate that same territory, a
goal which saw some brilliant successes in late 2022

in the Kharkiv and Kherson regions. However, another Ukrainian counteroffensive in 2023 failed and in early 2024 it seems that Ukrainian commanders decided to move to a different strategy: a campaign of long-term attrition designed to gain ground by grinding down Russian fighting capability. Over the ensuing year, Russian casualties and losses of materiel (vehicles, ammunition, etc) have skyrocketed, even as their forward momentum has slowed. It's still too early to say precisely how the war will end—but what we're seeing is a steady decline in Russian combat capabilities. By changing their objective from control of territory to attrition of the Russian army, Ukraine is leveraging one of Russia's historical weaknesses: its willingness to waste the lives of its own people.

Often, war is best fought when you don't play by your opponent's rules, and don't grab for the same prizes.

Some Relevant Considerations

N ow that we've covered some of the groundwork that gets laid leading up to a battle, let's discuss some of the important tactical considerations we often see coming into play in real historical battles.

Surprise

Remember that I defined surprise as one combatant's bid to deprive the other of the ability to plan? Being caught by surprise is what Cohen & Gooch define in *Military Misfortunes* as one of the three basic failures: Failure to Anticipate. Needless to say, the bigger the surprise, the more likely your chance of success.

Surprise can take many forms—it's more than just the classic ambush or night attack, in which your opponent doesn't know that you're attacking at all. For instance, at the 1099 siege of Jerusalem, the First Crusaders initially constructed their siege tower facing the western wall, leading the defenders of the city to rush all their battle resources to that part of the wall. Overnight, however, the crusaders quietly took the siege tower apart again. When the sun rose, the defenders found the crusaders attacking a totally *different* part of the wall, where the defenders had far fewer resources.

Surprise, then, might involve the other side not knowing that you're going to attack them at all. Or if they know you're going to attack, they don't know when. Or they don't know where. Or they know when or where, but they don't know that you have reserves—or where you're keeping them.

Speaking of reserves, surprise is also not something that always happens at the very start of a battle. Right when your attack has settled down into dogged fighting, or is even beginning to retreat, is the ideal moment to surprise the enemy a second time by deploying reserves they don't know you have—or a whole ally, or an advanced tactic or weapon they aren't expecting. A surprise attack right when they've realised this

battle is going to be harder to win than they expected, like Aragorn's arrival on the Pelennor Fields, is often the decisive blow to break the enemy's resolve and
set them fleeing from the battlefield, spreading panic
among their other troops as they run.

Surprise isn't always achieved by the aggressor, either. In February 2022, the U.S. and its allies assessed
that the Ukrainian army would fall apart the moment
it faced a full-scale Russian invasion. The assessment
was wrong for a very specific reason: as far as the U
.S. could tell, Ukraine's Commander-in-Chief, General Valery Zaluzhny, had no battle plans whatsoever.
In fact, Zaluzhny not only had a plan; he was putting
it into action. Aware, however, that higher echelons
of Ukrainian politics had been infiltrated by Russian agents, he was not sharing the plan with *anyone*.
"I was afraid that we would lose the element of surprise," Zaluzhny recalled (as reported in Yaroslav Trofimov's book, *Our Enemies Will Vanish*). "We needed
the adversary to think that we are all deployed in our
usual bases, smoking grass, watching TV, and posting
on Facebook." Meanwhile, very quietly, Zaluzhny "ordered units to disperse around the country, ostensibly
for training exercises, and moved air-defense batteries
and military aircraft from their bases to new, hidden

locations," according to Trofimov. When the Russians began the invasion with a massive bombardment of Ukrainian airfields and other military infrastructure, much was saved by Zaluzhny's foresight. Ukraine's state of readiness itself was a complete surprise to the invaders.

Failing to make use of surprise is one of the most common mistakes I see authors making when it comes to writing battles, particularly when it comes to the good guys. Authors generally seem quite happy to allow the enemy to spring an ambush, but if the good guys are going to march out of their fortifications to face the enemy, they almost never go and conceal themselves sensibly in a nice bit of forest or behind a hillside. Instead, they tend to just stand about in the open countryside, offering a nice clear target to shoot at. I'm begging you: there's nothing wrong with letting your good characters be just a little cunning!

Terrain

Most people tend to think of battles in terms of two factors: men and weapons. However, if you're going to be planning a battle, whether as an author or in some other capacity—I won't judge—then you need to recognise that *terrain* is an equally important factor. Even if your characters can't pick the *ideal* place to fight, they can usually make the most of whatever advantages the landscape *does* offer. But very few writers seem to know about this. In fact, a failure to consider terrain is up there with a failure to make use of fortifications and surprise as one of the most common battle mistakes I see in film, TV, and written media.

You will need to put some serious thought into considering how both sides might use the landscape to their advantage—which means, of course, as a fantasy author, that you'll want to have your battlefield built out in detail. Hills and forests can be used to conceal your heroes' initial attack and whatever reserves or reinforcements they may have waiting in the wings. If they have to retreat, disappearing into the forest will help conceal them and complicate the enemy's retreat. Hills provide vantage points from which to use long-range weaponry—bows, artillery, spell-slinging wizards—against hapless enemies corralled into the valleys.

Roads can be blocked in order to funnel the enemy into a desired ambush. Marshes, rivers, cliffs, or coastlines can be used defensively to protect your flanks. Any side protected by a cliff or river is a side the enemy probably won't attack you from (unless they surprise you in turn with ships, bridges, dragons, etc). Or positioning yourself so that an enemy has to cross wetlands or deserts in order to fight you can deplete her strength, making her easier to fight once she actually reaches you. Terrain can also protect you while you deploy your troops for battle, buying time and complicating the enemy's advance.

During the February 2022 Battle of Voznesensk, elements of the Russian army approached this small Ukrainian town seeking a way across a strategic bridge on the Deadwater River. Since the open ground on the south bank of the river from which the Russians were approaching offered no good defensive positions, the Voznesensk defenders chose to rig the bridge to explode and then make their stand in the town itself. But when should the bridge be destroyed? The Ukrainian army commander wanted to allow part of the Russian column to cross the bridge and enter the town before exploding the bridge. The city of Voznesensk itself could then be turned into a death-trap for those unlucky enough to make it across the river. And the remaining Russian forces would be left crippled, incapable of mounting a new attack on the town. On the other hand, the town's mayor wanted the bridge

destroyed well before the Russians arrived, thereby en-
suring the safety of civilians in the town who might
otherwise get caught in the fighting. In any case, it was
the mayor who got his wish when a technical fault made
it necessary for a sapper to blow up the bridge early
rather than risk failing to destroy it at all. In any case,
such was the intensity of the Ukrainian defence that the
Russians were soon forced to retreat anyway. (Andrew
Harding's book on this battle, *A Small, Stubborn Town:
Life, Death, and Defiance in Ukraine* is a quick read and
an excellent study of how terrain affects war even in the
21st century).

Far north of Voznesensk, around the same time, the
Battle of Kyiv also hinged on Russian attempts to cross
the Irpin River towards the city. With the original Irpin
bridge blown up by Ukrainian sappers, the Russians
managed to lay a small pontoon bridge as a replace-
ment. Facing the prospect of an imminent Russian
crossing in force, Ukrainian Ground Forces commander Oleksandr Syrsky sent reconnaissance units behind
enemy lines to the spot at which the Irpin flows into
the Kyiv Sea, a large man-made reservoir north of the
city, where sluice pumps facilitated the flow of water
into the reservoir. With the sluices sabotaged, the Irpin
swelled—slowly at first, and then quickly. The banks of

the river turned into a marsh, and then a shallow lake more than a mile wide. Soon, the Russians who had already crossed the river via pontoon into the suburb of Moschun were cut off, and a further crossing was impossible.

The Russian army of 2022 was not the first army to fall into such a trap. Nearly eight centuries had passed since the Seventh Crusade had been cut off and forced to surrender in Egypt by the annual flooding of the Nile—yet although the Russians possessed modern tanks, artillery, drones, and firearms, they, too, were halted by a flooding river just the same.

Terrain isn't the only element of the landscape that might be used to your advantage: weather might also play a role. If the wind is in your favour, you might light bonfires to belch smoke into your enemies' eyes and conceal your movements. Or, in naval contexts, wind might allow you to send blazing fireships into the opponent's fleet, as the English defenders did to the Spanish Armada in 1588. The cover of night can be helpful in planning a surprise attack—or the bright sun might throw a photosensitive fantasy race into disarray, like the Orcs at the Battle of Helm's Deep. Rain might soften the ground, bogging down your opponent's heavy cavalry or artillery; in settings with gunpowder, it might spoil an opponent's gunpowder, as happened during the 1757 Battle of Plassey, a decisive victory for the East India Company.

Many authors never even think about the vital importance of terrain in battle. The moment you do, you'll be far ahead of them.

Reserves

Reserves are a group, often of heavy cavalry, that is held back specifically to help out at a decisive moment—for example when something goes wrong. If your attack is flagging, or if your defence looks like breaking, your reserves are there to reinforce the tiring men. If you need to retreat, your reserves will probably be the one thing standing between your fleeing men and the enemy trying to massacre them. In the 2005 *The Lion, the Witch, and the Wardrobe* movie, King Peter makes the mistake of ordering a retreat not only without any reserves to cover the retreat, but also without a defensible position to fall back *to.* The result is a realistic and completely predictable and avoidable bloodbath.

On the other hand, if something goes right, reserves are more helpful still. They might launch a surprise attack right as the battle has stagnated, breaking the enemy's resolve. In the best case scenario, if all goes well and the enemy begins to flee, your reserves can join the chase to capture the enemy camp and ensure they don't get the chance to rally, regroup and counterattack.

The Lake Battle of Antioch

Surprise, terrain, and reserves all played a key role during the 1098 Lake Battle during the First Crusade's siege of Antioch. The crusaders had barely 700 mounted knights, many of them forced to ride on pack animals or oxen rather than horses, and they were facing an approaching Turkish army of around 12,000. Since the crusaders had no fortifications to hide behind and no hope of reinforcements to rescue them, this was one of the situations in which it made more sense to adopt an aggressive strategy than to fight defensively. Accordingly, they chose to march out and ambush their opponents rather than allow themselves to be caught like sitting ducks outside the walls of Antioch. The crusaders' most brilliant commander, Bohemond of Taranto, chose a place for his ambush where a series of small hills would conceal his entire force from the road along which the relieving army marched. When the larger army approached, five of Bohemond's six squadrons of knights emerged from the hills and tore into the Turkish vanguard. Caught by surprise, the opponent's vanguard was forced to retreat, blundering into the Turkish army behind and disrupting their efforts to deploy for battle. The fighting quickly bogged down, however, with the shock of the first charge dissipating and the majority of the Turkish army refusing to flee. That's when Bohemond unleashed his sixth and largest squadron in a charge that threw the entire, much larger, Turkish army into a chaotic rout. His knights pursued the Turks for miles. So complete was the victory that a fortress some miles away to which many of the Turks fled for refuge actually surrendered when they saw the pursuing crusaders.

Supply Lines

Supply lines are absolutely essential to any army. In hostile territory, everything your army uses—food, water, medicines, weapons, horses, vehicles, spare parts, reinforcements, secure routes of communication and retreat—will all need to be sourced and protected. The further into enemy territory you move, the longer and more vulnerable those supply lines will become, and the easier your opponent will find it to cut you off and encircle you (which is one reason you should always be careful about advancing too far, too fast). Even in your own territory, neglecting supply can spell death. At the battle of Hattin in 1187, Guy of Lusignan made the crucial mistake of leaving his water supply and marching his army across miles of arid terrain to confront his opponent, Saladin. By the time the battle actually came about, Guy's army was exhausted, parched, and ill with heatstroke.

Breaking supply lines is a basic siege tactic: without fresh food, new blood, or weapons, the besieged will be forced to negotiate or surrender. Accordingly, if you or your friends are being besieged, resupplying will be a major priority for you. If food and reinforcements can be shipped in somehow—say, by sea—then you might resist the siege indefinitely. This is how the 1189-1191 siege of Acre was able to continue for three long years, forced to surrender only once the Third Crusade succeeded in breaking down the walls of the city with catapults.

Urban Warfare (Don't)

Urban warfare—street-to-street or house-to-house fighting in a cityscape—is something we normally think of in the context of modern warfare, for instance during the WWII Battle of Stalingrad. I'm not going to say much about this—it has always been uniquely dangerous and difficult—except: Don't.

Today, urban warfare is a specialised field precisely because a cityscape provides such a complex and defensible battlefield—which is why the Ukrainian army wanted to lure the Russians into urban warfare during the Battle of Voznesensk. In medieval times, urban warfare was just as dangerous, except that no one was specially trained to survive it.

And yet if I had five dollars for every time I've read a fantasy novel in which some bold leader with a small, handpicked band infiltrates a city and successfully captures it, I'd be in a position to fund a whole funeral for Robert, Count of Artois, and three hundred elite knights during the Seventh Crusade. During the Battle of Mansourah in Egypt, Count Robert and a detachment of Templars and English successfully overwhelmed an Egyptian outpost. The outpost garrison fled towards the city of Mansourah, Count Robert charged gallantly after them, and the Templars charged after him. An eyewitness of the battle, John of Joinville, described what happened next:

> At this the Templars, thinking they would be shamed if they let the Comte d'Artois get in front of them, struck spurs into their horses and rushed headlong in pursuit of the Turks, who fled before them, right through the town of Mansourah and on into the fields beyond towards Cairo. When our men tried to return, the Turks in Mansourah threw great beams and blocks of wood down on them as they passed through the streets, which were very narrow. The Comte d'Artois was killed there, together with Raoul de Coucy and so many other knights that the number of dead was estimated at three hundred.

This was not the only time something like this happened. As a rule, one of the easiest ways to die in medieval warfare is to get trapped in narrow city streets. And it wasn't as though Artois was basically alone, either—he had three *hundred* men with him. There's a reason why street fighting normally only happened at the end of a successful siege, and relied on the whole army to overwhelm the townspeople's resistance. If you're going to have a small force infiltrate a medieval city, be mindful that this is more or less a suicide mission.

Mistakes

Just like Count Robert of Artois, people are likely to make mistakes in the heat of battle. These mistakes may be made by commanders, or they may be made by the rank and file. They may be made freely or forced by an opponent, as happened in the Great Battle of Antioch in 1098, when Bohemond of Taranto's battle plan was designed to offer his opposing general a series of brutal dilemmas. They may be the result of overconfidence, the "fog of war", or simple greed.

At the Battle of Jaffa during the Third Crusade, the entire force of the Knights Hospitaller committed a mistake when they prematurely charged their opponent, opening the battle before the best opportunity

had come. Only Richard Coeur-de-Lion's able gener-
alship saved the situation, instantly sending the rest of
the army straight off after the Hospitallers.

During the seventeenth-century English Civil War,
the royalist cavalry under Prince Rupert developed a
bad habit of putting loot above victory. Again and again
they would charge the Parliamentarian troops and scat-
ter them—then ride off to steal things from the Parliamentarian baggage train. While these noble cavaliers
were grabbing everything valuable they could find, the
Parliamentarian troops would rally, regroup, and then
deal the royalists a shattering surprise attack culminat-
ing in a terrible defeat.

A cunning general may see the opportunity to force
an error, either by presenting his opponent with a tricky
dilemma, or simply by taunting them. At Hattin in
1187, with the crusaders fainting for lack of water, Saladin sought to force an error by driving water-laden
camels within view of the crusader army. When the
crusaders failed to charge after the camels, he had his
men slash the water-skins, releasing the priceless water
into the dust. Even without eliciting the suicidal charge
Saladin wanted, the impact on the crusaders' morale
must have been significant.

A clever general will be aware of the mistakes he or his men are likely to make, and act to mitigate them ahead of time. Richard Coeur-de-lion managed this at the battle of Jaffa, by having his whole army deployed into battle array so that when the Hospitallers made their early charge, he was ready to dispatch the rest of the army to back them up.

Adaptability

It's an axiom of warfare that no plan survives contact with the enemy. Therefore, one of the most important skills of a good commander is the ability to adapt. When the enemy achieves surprise, when circumstances unexpectedly turn against you—or even when things go unexpectedly well—a good commander must be able to adapt on the fly, circumvent the crisis, or take advantage of opportunities.

Adaptability isn't just something for commanders, however. Remember what I said earlier, about Soviet versus NATO "mission command" styles? The rank and file also need to be able to take initiative when necessary, to press an advantage or to source important supplies for themselves, without waiting to be told. Communication lines are another absolutely vital element of the ability to adapt. The men need to be able to communicate with their commander about conditions on the ground, and the commander needs to be able to inform her men about the new plans. It doesn't matter how smart the commander is—if she's unaware of what's happening in the battle, or if she's unable to communicate with her men, she's not going to be able to observe what's happening or adapt her plans.

Fog of War

The term "fog of war" refers to the element of uncertainty in war. A classic definition of fog of war from an 1896 book by Sir Augustus Lonsdale Hale is, "the state of ignorance in which commanders frequently find themselves as regards the real strength and position, not only of their foes, but also of their friends." In combat, soldiers may become confused and disoriented as to what is happening and where. At higher levels, this is why military intelligence is so vital to provide information to commanders about the movement, location, and numbers of the enemy.

Fog of war could be particularly crippling in pre-modern battle, when notions of honour demanded that commanders "lead from the front", being present on the front lines with their men. Under these conditions, battle leaders could become lost in the fog of war on both levels, losing the battle simply because they weren't fully aware of what was going on in other parts of the field. These days, with ever more sophisticated surveillance and reconnaissance technology, such as drones and satellite imagery, the fog of war is being dissipated to an extent never before experienced in human history...but still not entirely.

Failure to Adapt is another of Cohen & Gooch's three basic military failures in *Military Misfortunes*. Their example of this failure is the British campaign at Gallipoli in World War I. Launched on an arid peninsula, the English troops failed to show initiative in finding and securing water sources for themselves. (The Australians, who'd come from a much drier and more inhospitable climate, did not commit the same mistake). Worse still, the commander of the expedition remained snugly on his battleship some way out at sea, incapable of communicating with his men on the ground except by boat. The expedition had a real chance to secure the peninsula, but failed to do so because their window of opportunity passed while the English troops were sitting around, waiting for orders from their distant commander.

The Evil Overlord who punishes any failure with death has been a standard trope of the fantasy genre at least as far back as Darth Vader. In reality, such a commander would be peculiarly vulnerable to this kind of failure because his men would be too frightened of him to take the initiative and possibly even to communicate honestly with him about conditions on the battlefield. On the other hand, the knightly hero leading the battle from the front, charging about on his horse, is also going to be uniquely vulnerable because he'll be enmeshed in the fog of war as deeply as his men. Both these commanders will have trouble adapting to the changing face of the battle.

Categories of Warrior

L oosely speaking, on the actual battlefield, you'll probably have a few main sorts of fighter to command. I'm going to be massively reductive here and call them cavalry, infantry, archers, engineers, and air forces. But as we'll see, each of these categories is a pretty broad umbrella.

To sum up: cavalry are mounted and used to attack. Infantry go on foot and are used to defend or consolidate. Archers attack at a distance using ranged weapons. Related, engineers use specialised machinery against fortifications. And an air force bombards the enemy from the air, as well as conducting reconnaissance.

Cavalry

If you're writing medieval style war, your cavalry are going to be largely composed of classic, heavily armoured knights. The rich men of your society, trained since birth, they're heavily armoured, ride equally heavy war-horses, and their attack—a massed, tightly packed charge with couched lances—is going to be the most devastating blow struck on the battlefield. They can be counteracted by any number of things—for instance, soft and marshy ground, uphill slopes, or caltrops (spikes designed to injure the hooves). Arrows are a particular hazard for these troops, not because the arrows are likely to penetrate the knight's own armour, but because it's very difficult (and expensive) to put armour on a horse. The more horses can be killed ahead of the battle, the fewer mounted knights your opponent will have to deal with when the battle comes.

A massed cavalry charge can also be stopped cold by a determined line of infantry with long pikes and a shield wall. But they have to be determined, well-trained, and well-armed. Don't expect frightened peasants with pitchforks to accomplish this.

Light cavalry may also be useful when you need a force that is highly agile and quick-moving. At the battle of Hattin, Saladin used his light cavalry skirmishers to splendid effect harassing the crusader army on their march. Light cavalry has also historically been useful in an intelligence capacity, as scouts—able to cover long distances quickly, spy out the land, identify and count enemy forces, and then return to the commander with news.

What if you're not writing medieval style war? The role of cavalry used to be played in ancient war by horse-drawn chariots. Today, to the extent that anything has replaced the heavy cavalryman, battle tanks are arguably the modern equivalent used for charging and penetrating enemy lines. However, an end might be coming to those days. Just as the invention of the gun rendered plate armour obsolete, so modern drone usage is making tanks less effective. In Ukraine right now (May, 2025), both armies are trialling a more nimble, less costly form of cavalry assault—putting assault troopers on motorcycles rather than packing them into tanks or armoured personnel carriers. The verdict is out on whether this new tactic will actually be effective, but it's a fascinating example of how technological innovation drives military adaptation.

Infantry

The infantry fight on foot and are neither dashing nor romantic, but they get the job done. Infantrymen, like cavalry, may be heavily armed and highly trained (such as the medieval "sergeants"), or they may be a motley rabble mostly armed with pitchforks, but in either case they are probably the ones who will win your battle for you. Marching in a Greek phalanx, armed with pikes in a shield-wall, trained to perform complex firing manoeuvers with early modern muskets, or armed with man-portable anti-tank weapons today, infantry are slow-moving but uniquely difficult to overcome.

When the cavalry creates a breach, it's the infantry who are able to exploit it, advancing to clear the new ground of remaining enemy forces and hold it against counterattack. In medieval warfare, moreover, it is the infantry who protect the cavalry until the time comes for a massed cavalry charge. At the Battle of Dorylaeum on the First Crusade, with the crusader vanguard completely surrounded by a Turkish army, the knights dismounted to form a ring of immovable heavy infantry surrounding the camp—a wall which stood unbroken against repeated assault for hours on end until the rest of the crusaders could arrive to relieve their co-religion-ists. Later, the crusaders actually developed a marching formation in which the knights would ride at the centre of the column with the infantry lining their flanks, screening the precious warhorses from arrows and attacks even while they were on the move.

Ranged Attacks: Archers, Engineers, Air Forces

In pre-modern warfare, the most common ranged weapon on the battlefield is the arrow. But history has seen a proliferation of ranged weapons and those who use them—not just archers, but slingers, snipers, and artillerymen. In the fantasy genre, you might have wizards with fireballs or wraiths riding eldritch monsters. In the modern day, you have the factors of artillery and aerial bombardment.

Medieval archers could be light cavalrymen or infantry who took to the bow when ranged attacks were to be made. The development of the English longbow, which was less powerful than the crossbow but could be fired at a much more rapid rate, had a powerful effect during the Hundred Years' War between England and France in the 14th century. A corps of longbowmen could throw back a charge of mounted knights. Archers also need to be highly trained, which is why for centuries in medieval and Renaissance England adult men were required to spend a certain amount of time shooting every week. This innovation was not copied in France partly because the nobility feared what might happen if the disgruntled peasants, who revolted several times that century, should do so as a corps of determined and highly-trained archers.

Bows, like all ranged weapons, are most useful in any other phase of battle than the general melee: if you're shooting from a distance, you don't want to be hitting your own men. Arrows and other ranged weapons up to and including the air force are traditionally used to soften up the enemy before an attack, as well as to harass them between attacks—all with the intent to create weaknesses which a heavy cavalry charge can then exploit.

In medieval times, you also had engineers—skilled, highly-trained professionals who built, calibrated, and managed siege engines, from trebuchets to mangonels and many others. Some of these machines are still only known to us by name, so that we're still not quite sure how they looked or what they did. I've chosen to lump them in with archers because these were ranged weapons and in some ways were the forerunners of modern heavy artillery. However, there are three distinctives that set the engineering corps aside from archers.

First, engineers are notable because they were among the first professional branch to develop in armies. While archers, infantrymen, and even knights could be recruited from among the general population, it took a very high level of specialised knowledge to become a siege engineer. As an engineer, you were a middle-class craftsman, a career soldier and not a recruit from among the general population.

Second, engineers were not just important in managing siege engines. Their other main function was in preparing roads, bridges, camp-sites, and fortifications for the army on the march. Having an army is all very well, but it must be able to move to where it is needed and be protected once it gets there. During a field battle, therefore, engineers will more likely be found siting and securing the camp, rather than participating in combat.

Third, when found in combat, engineers were useful particularly during sieges. A siege engine is designed to be used against a stationary, not a moving target, and most engines were much too heavy and unwieldy to be easily manoeuvered on the battlefield. Moreover, aiming a siege engine was a complicated process that involved making several bracketing shots and narrowing down aim from there. This is why one of the most ridiculous things I've ever seen in an onscreen battle happens in Disney's live-action *Mulan,* when the light cavalry nomad army brings *siege engines* to the battlefield for use against quickly-moving mounted opponents!!!

The invention of the cannon led ultimately to the modern field gun, which overcame all these drawbacks. Modern artillery is mobile and quickly aimed—an ideal battlefield weapon—and thus became not just the descendant of the medieval engineer, but also of the medieval archer.

Aerial bombardment is something we think of today strictly as a twentieth-century innovation, but since we're writing fantasy, it's worth saying something about this factor too. I'm not fully familiar with the use of air forces in modern combined arms combat, but one of the most notable uses of an air force in the modern era comes at the beginning of an attack, with a heavy aerial bombardment of the opponent's military infrastructure, fortresses and encampments. If you're working with, say, dragons or gryphons or the like, this is the way you'll probably be using them. Air forces are also responsible for flying reconnaissance missions, though this may be less important in an age of satellite imagery. Finally, during a battle, air forces may have a similar role to other ranged weapons, bombarding the enemy between charges.

Combined Arms Combat

Combined arms seeks to integrate the different combat arms of a military—i.e., air forces, armoured vehicles, infantry, or navy—to achieve mutually supportive effects. For instance, modern urban warfare in which infantry and armoured tanks act to support each other, or an invasion in which the air force or navy works in concert with the men on the ground. Today, cyber warfare and drone forces add new elements to combined arms combat.

While historical medieval use of combined arms might be limited to making sure that the cavalry, infantry, and archers are all working smoothly together, a fantasy battle may benefit from an understanding of modern combined arms doctrine, depending on how your magical elements affect your battlefield.

Now Add Magic

If the idea of adding fantasy elements seems intimidating in addition to the challenge of writing a credible battle...don't be scared. The easiest thing to do is to consider how your worldbuilding can best map on to some real-life element of war. For instance, let's assume that we're writing something set in Tolkien's Middle Earth. Dragons, as I've mentioned, take the place of air forces: in *The Silmarillion,* Morgoth begins the Dagor Bragollach battle with an aerial bombardment by his dragons. This takes the Elves, who have no comparable weapons, utterly by surprise. Similarly, Tolkien's palantiri, or seeing-stones, are obviously most useful (and most dangerous) as intelligence-gathering tools. And while night attacks are standard surprise practice in human warfare, when you're fighting Orcs with their intolerance of sunlight, daytime attacks are most helpful.

This, of course, is also why military history beyond medieval history can be helpful, even if you're writing medieval-style fantasy: because magic can supply what in the real world came about as a result of technological advances. If you have a fantasy air force, you'll want to know some things about how modern air forces are used.

Phases of Battle

A t this point, I'm going to undertake the deeply silly exercise of describing the "average" medieval battle. As mentioned previously, there's really no such thing, and it's mostly by studying real battles that you'll come to appreciate the full breadth of stuff that can go wrong (or right, depending on your point of view) on the battlefield.

But there are an awful lot of commonalities to the way that battles unfold, especially in the context of medieval warfare. So I'll try to describe them here. I hasten to add that this is my very own taxonomy, created in an odd five minutes of the afternoon, and not something that I learned from someone who actually knows.

Phase 1: Setup

In this phase, the opponents have mustered their forces, laid their plans, identified their objectives, and are now trying to manoeuvre each other into a disadvantaged position. Advantage or disadvantage most commonly has to do with terrain, although psychological, supply, or health factors may also be relevant.

An unsung but important factor at this phase is the use of scouts—usually light cavalry—or other intelligence-gathering tools, from spies to air reconnaissance aircraft or drones. You cannot gauge the right moment for attack if you do not know what the enemy is doing, so it is vital to send out small, mobile forces to probe the enemy and bring back intelligence. Sending out scouts will not always prevent you falling into an ambush or other surprise attack, but failing to do so will guarantee it.

At the 1187 battle of Hattin, the setup phase involved the two opposing armies facing off across approximately twenty-four kilometres (fifteen miles) of arid summer landscape in the northern Kingdom of Jerusalem. Temperatures in Galilee in July reach an average high of 34° Celsius (94° Fahrenheit), and a recent drought had depleted water supplies across the kingdom. Both armies were aware that whoever took the initiative by crossing those waterless hills would be at a disadvantage once they reached the other side. Both commanders were under strong political pressure to give battle in order to justify their rule to their followers. For several days they waited, watched each other, and planned. At one point, Saladin advanced halfway to King Guy's camp in an effort to draw him into battle, but Guy did not respond and Saladin was forced to retreat again. His next move was to strike northwards towards the city of Tiberias on the shore of the Sea of Galilee. In a one-day siege he seized the city from its lady, who was forced to withdraw with her men to the citadel. King Guy was now honour-bound to come to his vassal's aid, to succour a lady, and to prevent Saladin securing his grip in the kingdom. In the small hours of 3 July, 1187, Guy gave the order for his army to leave camp and march towards Tiberias: Saladin had successfully manoeuvred Guy into giving battle on Saladin's terms.

Phase 2: Distance Contact

Most of the time, hostilities begin with ranged attacks: slings, arrows and other kinds of missile, or (these days) aerial, artillery, and drone bombardment. The main purpose of this is to soften the enemy's morale, discipline, and defensive arrangements in order to create an opportunity for attack.

During the battle of Hattin, the distance phase began around noon on the first day, while King Guy's army was already on the march. He and his men had covered the first half of the march to Tiberias, stopping around mid-morning to drink the (insufficient) water at the wells of Turan. They were travelling in three squadrons, each consisting of knights mounted on rounceys (all-purpose riding horses) with their warhorses led behind, protected by a screen of infantry. As this column climbed into the Galilean hills, Saladin ordered his mounted archers and light cavalry skirmishers into action. Harassed by constant volleys of arrows in blistering heat, the crusaders began steadily losing horses and men.

It was probably Guy's plan to get to grips with his opponent that first day, but he had underestimated Saladin's numbers, and the pressure from skirmishers became so unrelenting that Guy was simply unable to deploy his men for battle. With the column still holding formation, securely protected by the infantry, however, there was little Saladin could do to defeat the crusaders. Still, time was on Saladin's side. As the punishing day wore on, Saladin continued to harry the crusaders, slowing their march to a crawl. By the time darkness fell, Guy's men were stranded on a high plateau south of the village of Hattin, still several miles short of Tiberias and the precious water of its lake.

When morning came, Saladin drew back just far enough to allow Guy to properly deploy his knights on their warhorses. Nearly twenty-four hours away from their last water source, and exhausted from the previous day's punishing march, Guy's army was now at a distinct disadvantage.

Deployment

In popular media, the only thing required to begin a battle is a big-name star in a shiny hat yelling "Charge!" at the top of her voice. In reality, soldiers take some getting ready. They need to be equipped, organised into the desired formation, and moved to the location where they're needed. For instance, medieval knights normally would not ride their war-horses on a long-distance march. This may surprise those of us who've grown up on Hollywood imagery of fully armed knights galloping hell-for-leather across miles of rolling grasslands in order to reach a distant battle on time, but horses are not motorcycles. They get tired and may even collapse from exhaustion. Knights would ride ordinary horses, "rounceys", until battle was clearly imminent. Only then would their squires prepare the great warhorses, "destriers", whose precious energy needed to be saved for the charge.

As you can imagine, deployment is not something you can do while people are actively trying to kill you. And needless to say, anything you can do on the battlefield to hinder or confuse your opponent's deployment is a point to you.

Phase 3: Full Contact

It's true that some battles *do* skip one or both of the first two phases—like in the Foraging Battle of Antioch, when two forces bumped into each other by complete surprise in the middle of nowhere. Remember that any successful use of surprise is designed to deny the enemy time to plan, manoeuvre, and weaken you.

The full contact phase begins—in medieval style war, at least—when someone makes a charge with their heaviest troops. Usually, this means armoured knights. It also usually means that the other side needs to forget about attacking for a bit and focus on the defence until they see their own opportunity for a charge. That is: rather than the Hollywood trope of a two-sided charge resulting in a confused melee, medieval war was more often an ebb-and-flow pattern of charge and retreat, charge and retreat by the attacking side; with defenders able to mount a countercharge only once the attackers' first energy is depleted.

During this phase, the aim is to secure a victory—but what does victory look like? The military theorist Clausewitz defined victory in war as a condition in which the enemy's ability to enter battle, resist or resume hostilities is destroyed. My own definition of victory in battle would be similar—the destruction of the opponent's ability and will to continue fighting. In battle, this might come about by capturing or killing the enemy's leader, forcing him to surrender, or by sending his rank and file into a panic. The latter is particularly effective because it can happen suddenly and devastatingly. In the best case, sowing panic among your opponent's vanguard may result in the complete physical and psychological breakdown of the entire rest of his army, as those who flee in a panic plough through other formations, breaking them up and entangling them in a general rout.

Vanguard and Rearguard

The term "vanguard" originates with the medieval French term "avant-garde", or "advance guard". Medieval armies were customarily deployed into three parts or "wards": the Van, the Main, and the Rear.

A vanguard generally forms the leading part of an army on the march. In medieval and Renaissance warfare, a vanguard included scouts who might ride out to spy on enemy movements; negotiators and heralds who might bear messages and summon enemy towns or castles to surrender; engineers whose task it was to ensure that the roads were clear; and harbingers who could locate lodgings for the army each night.

The term "rearguard" originates with the medieval French term "rere-garde", or "rearward". As you might expect, the rearguard formed the hindmost part of an army on the march.

The task of the rearguard is to protect the rest of the army from attack in the rear, and sometimes also to secure lines of communication or occupy tactically important fortifications or terrain. The rearguard may also be used offensively, traditionally being the people who conduct spoiling attacks--attacks with the limited objective of disrupting the enemy's plans, for instance their attempts to deploy their troops into battle array.

The rearguard is often most important in protecting the army if it is forced to retreat--these are the guys who keep their faces and weapons pointed at the enemy while the main force falls back to regroup. Rearguards are usually formed with an eye to manoeuvrability, so they may be cavalry, tanks or mobile infantry.

At the battle of Hattin, the full contact fighting began on the morning of the second day, 4 July, a few hours after Guy had deployed his men into combat readiness and set off, now aiming north towards the water springs at the village of Hattin. Saladin now withdrew brought up his main army, though he continued to hold off from a massed charge, probing the crusaders with archers and light cavalry skirmishers. It was at this point that he tried to tempt Guy's knights into a reckless charge by parading camels bearing water-skins before them. He also had his auxiliaries light brushfires downwind of the crusaders, smothering them in smoke to exacerbate their thirst.

Crusader battle etiquette gave the lord of the county where the fighting was taking place the command of the vanguard and the right of first charge, so it was Count Raymond, whose lady was awaiting relief in the citadel of Tiberias, who made the first charge towards the north. What happened next is unclear. Raymond was likely being opposed only by the light cavalry of Saladin's vanguard, and it was their practice not to try withstanding a crusader massed charge. They permitted the count and his knights to pierce their ranks, then closed up again behind him. Count Raymond then decided, rightly or wrongly, that it was useless trying to rejoin the main body of the army. He led his knights away past Hattin to safety at the city of Tyre, and died soon after under suspicion of treason. The vanguard infantry, who of course had been left behind in the charge, now sought refuge on one of two hilltops over-looking the distant Lake Tiberias and refused to come down.

Meanwhile, the two remaining squadrons—the main body of the army commanded by King Guy, and the rearguard—had also begun making charges against Saladin's troops. As the day wore on, the two crusader squadrons became separated from each other, and Guy was pushed slowly but surely up the second of the two hilltops. This was the logical place for his last stand: the sheer cliffs at his back would have protected his rear, while the slope at his front added power to his own cavalry charges and impeded those of the opponent. This advantage did not succeed in changing the course of the battle. It was the battered rearguard, some way off, which next saw the opportunity to escape the bat-tlefield towards the south, as Saladin's army concen-trated all its force on the king's squadron. Guy's men, however, were trapped between the precipice at their backs and their opponents at their front. They had no choice but to go on fighting in a series of desperate charges that nearly reached Saladin's own standard and might have won them the battle had they been able to kill or capture him. Ultimately, however, Saladin's army prevailed, capturing the king's tent and standards, signalling to the remaining crusaders that the battle was lost.

Before we move on to the final phase of battle, here's a list of things it may be helpful to remember when writing the full-contact phase.

- The best attack is usually heavy cavalry; the best defence (in the absence of geographical obstacles or fortifications such as walls and trenches) is usually heavy infantry.

- Defenders always have the advantage, especially when there's been time to prepare fortifications. It takes many more men to capture a position than it does to defend it.

- Never forget the way terrain may shape the battle and be used to disadvantage or benefit your protagonists.

- Ranged weapons become less useful during the full contact phase, as you don't want your archers hitting your own people. They may still release volleys of arrows between charges, or they may transition to fighting as light infantry.

- The goal of the attackers is generally to break the enemy lines and their will to fight. The goal of the defenders is to stop or absorb all attacks for long enough to mount a counter-attack.

- When reserves come into play, it's usually sometime during the full contact phase—often the unexpected arrival of reserves becomes the final blow that breaks the enemy's will to fight.

Phase 4: Cleanup

Often a battle ends when one opponent delivers that one decisive strike that disrupts the enemy's remaining formations, sows panic, and sends the combatants fleeing from the battlefield in disarray. Sometimes this comes about dramatically, when reserves are thrown in, but sometimes it's also just the result of hard and determined fighting that ultimately grinds down the enemy's will to fight.

In the best case scenario, when panic begins and people start running from the battlefield, they'll break up their own remaining formations and even reserves in their haste to get away, and a pursuit can ensue. This is what happened at the decisive moment of the 1098 Battle of Antioch. However, even at this point, luck can change. A determined force of reserves, or a single formation that manages to keep its nerve, might avoid the panic of the rout and cover the retreat, allowing their army to fall back in good order to a predetermined rallying point from which they might yet stage a comeback. Even if the enemy is retreating in disarray, a wise commander won't allow himself to become complacent. Many a self-declared "winner" has got carried away with looting the now vulnerable enemy camp, only to have the enemy regroup, return, and destroy them. A too-bold pursuit can also trick the victors into over-extending themselves, as happened to Count Robert of Artois at Masourah. Once the enemy begins to collapse, it's important to remain vigilant for more trouble, while still identifying and exploiting any unexpected opportunities provided by the rout.

Not all battles end with wholesale panic and a pursuit. They might also end after a grinding battle, with the final few detachments of the enemy fought to standstill and either captured or killed. In such a case, whoever means to run away has probably already done so by the time the last holdouts are subdued.

At the battle of Hattin, as we have seen, the knights of the vanguard and of the rearguard made their escape while King Guy, who had been surrounded, battled on. It took the capture of the king, together with his standards, to signal the final defeat. At this point, some of the remaining army of Jerusalem may have been able to escape, but the majority would have been killed or captured. Of those who were captured, some—the Templars and Hospitallers, as well as light cavalrymen recruited from among the indigenous Christians of Palestine—were slaughtered. In the case of the latter, presumably Saladin felt he could not rule the kingdom if the indigenous population was willing to take up arms against him, and therefore felt the need to make an example of them. Nevertheless in any medieval war, a profit could usually be made from prisoners, so killing them did not usually make sense. The high-status prisoners were held to ransom, and many of the rank and file were sold into slavery. By nightfall, Saladin had secured the battlefield, shattered the army of Jerusalem, captured almost the entirety of the kingdom's leadership, and begun a victorious campaign that would end only with the arrival of the Third Crusade several years later. The risk of a pitched field battle had paid off.

I hope that this description of the nonexistent "average battle", in tandem with an account of a real battle, helps you to see not just how many battles do unfold, but also how specific, individual circumstances can shape individual battles. Again, it's by studying individual real battles that you'll begin to appreciate the true breadth of things that can happen in war—and how, even though much of the Hollywood picture of war is hilariously inaccurate, the reality is often just as dramatic.

Now Write the Battle

Understanding battles is one thing—writing them is another sort of challenge entirely. I've now written about more battles than I can count, and the main challenge, as I see it, is that battles are large-scale affairs in which the destinies of whole cities or kingdoms are at stake. These are big stakes, but they are not *personal* stakes, and you cannot make a story out of something with no personal stakes. Your readers are hopefully not reading your book for the military strategy; they're reading it for the characters. They're going to experience the battle through your characters' experience, which means that every time you sit down to write the battle, the challenge is twofold. First, you want to provide an accessible way for the audience to perceive what is going on. Second, you'll want this perspective to remain closely and very personally connected to your characters. In this, you will be limited by factors such as your book's point of view, your audience's expectations, and the amount of "screentime" the battle is supposed to take up.

You have a few choices to make here—and at one time or another, I've made most of them, because I've written a heck of a lot of battle scenes.

As regards point of view, you can either write a battle scene in one character's POV, or several. Writing in one character's POV can be very limiting, because one person in a battle is usually thoroughly caught up in the fog of war, even if they're a high-ranking commander. They don't know exactly what's going on, and as you should be beginning to grasp by now, no matter what role they take in the battle—commander or grunt, cavalry or infantry—they'll be spending long periods of time waiting for their moment to fight, rather than actually fighting.

On the other hand, a single POV can be helpful if you're vague on the actual battle tactics. A single POV shrouded in fog of war can't be expected to know what's going on or why, so you can just narrow in on their chaotic experience and let the broader picture remain vague to great dramatic effect.

By contrast, writing in several POVs (my own preference) will allow you to depict the battle from several angles, clearly explain the battle plan, and broaden the scale of your story to something that feels more epic. Multiple POVs disguise time skips and allow the narrative to keep moving to a part of the battlefield where something is happening, rather than remaining stationary with one character who may spend much of the battle waiting to be deployed. Of course, in the sort of book that gets described as "sweeping" or "epic", you'll probably already have several POV characters to follow through the battle—so try to diversify their roles and locations during the battle in order to give the fullest picture.

What about using omniscient narration to give the readers a birds-eye view of the battle? The pitfall here is that it becomes easy to step too far away from the characters and their experience of the battle, to the extent that the story begins to read like non-fiction. If using omniscient narration, I would be very, very careful not to let any part of the story get detached from the characters.

When it comes to how long a battle scene should go on for...that depends on your goals for your book, as well as your narration style. I've written battles that go on for chapters, spanning two, three or even more points of view, with the aim of describing the battle in detail. (In *The House of Mourning,* I spend a big chunk of the middle of the book purely on Hattin). But most stories aren't designed this way, and you don't want to give your readers more military details than they want.

Condensing a major battle into one chapter is harder, because a battle is *such* a big event. If you wish to do something like that, however, consider the way CS Lewis does it in *The Lion, the Witch, and the Wardrobe.* He deals with that battle very briefly. The setup phase is suggested in earlier chapters when we briefly see Aslan advising High King Peter on how he should command his army in the coming battle. Then, as the White Witch marches off to begin the battle, the story sticks with Lucy and Susan as they mourn over Aslan, receive him back from the dead, and set off to liberate the Witch's captives. They reach the battle with Aslan only as surprise reserves, a surprise which quickly routs the enemy and signals the end of the full contact phase and the beginning of the cleanup phase. By focusing only on the decisive turning-point of the battle, Lewis gives us all the drama and none of the tedium, while remaining focused on the emotional heart of his story.

Which leads to my final point: do remember that the battle itself is only ever part of the backdrop. Your individual characters and their individual journeys—*that's* your real story.

Writing highly accurate historical fiction in which at least one and often more battles occur in each book has forced me to be selective about which battles to write. Just because the book calls for a big climactic battle—whether because that's what's supposed to happen at the end of a fantasy book, or because the history inspiring your story happens to revolve around a pivotal battle—doesn't mean that your characters need to actually *be* at the big battle. In one of my biggest battle sequences—the 1098 battle of Antioch in *A Conspiracy of Prophets*—I ended up not putting any of my POV characters on the battlefield at all. Instead, I wove together three narratives. One of them is a series of flashbacks to the meeting where the commander, Bohemond, explains his plans for the battle. This is interspersed with the POV of one of the participants of that meeting, who is watching the battle from a vantage-point atop a mountain which it is his job to guard. This gives a birds-eye view, with commentary provided via the flashbacks for why the battle unfolds the way it does. Finally, my most important characters spend the entire battle in the nearby city having their very own climactic confrontation, the one involving magic. The battle is important to the characters—all of them are at risk of dying if it goes badly—but in the end it wasn't the real heart of the story, and I didn't need to put my characters through it in order to make it an epic part of the backdrop.

Conclusion

I will go to my grave maintaining that real battles are amazingly dramatic, and that you don't need to make up bad tactics for the sake of the story. That said, real life is not fiction and vice versa. I've spent a lot of time in this little book being smart-alecky about how popular media gets battles wrong, but in the end, the main thing is (usually) not to be totally accurate. The point is to tell a story that *works*. That's why you might not fully depict every battle, or why even with a massive climactic battle going on, your characters might not need to be part of it.

My goal in this book is not that you write battles which are impeccably realistic—although that would make me a very, very happy reader. My goal is to demystify battles for you, so that writing them isn't an endless headache. Your goal as an author doesn't need to be impeccable realism, either. You just need to be credible enough to enable the audience's suspension of disbelief. And, in order to do that, you just need to know a little bit more than your average reader, whoever they may be.

I hope this little book has helped you to do just that—and if it's been helpful to you, I do encourage you to check out some of the more serious books I mentioned along the way. I've included a list of them on the next page.

Now: go and write that battle!

S.D.G

Acknowledgements and Further Reading

Thank you for reading! If you found this book helpful, please consider mentioning it to a friend, or leaving a review with your favoured retailer.

I also want to thank those who helped bring this book to fruition. My writing besties at The Thing With Feathers, particularly Joanna Ruth Meyer—thank you for not just inspiring this book, but also cheering me on as I dithered over whether the world really needed an entire book of me being salty about military tactics. Huge thanks are also due to Rob Wallace for being willing to put his military history degree to use checking my work. I hope I have now done justice to the engineers.

Needless to say, any errors that remain are entirely my own.

Should you wish to read further on this topic, here are the titles referenced in the text of this short book:

Medicine in the Crusades: Warfare, Wounds, and the Medieval Surgeon, by Piers D. Mitchell

Western Warfare in the Age of the Crusades, 1000–1300, and *Victory in the East: A Military History of the First Crusade,* by John France

The Origins of War: From the Stone Age to Alexander the Great, by Arther Ferrill

De Rei Militari, by Vegetius

On War, by Carl von Clausewitz

Strategy, by BH Liddell Hart

Military Misfortunes, by Cohen and Gooch

The Osprey Campaign Series

The Anarchy: The East India Company, Corporate Violence, and the Pillage of an Empire, by William Dalrymple

Our Enemies Will Vanish: The Russian Invasion and Ukraine's War of Independence, by Yaroslav Trofimov

The Lord of the Rings and *The Silmarillion,* by JRR Tolkien

A Small, Stubborn Town: Life, Death, and Defiance in Ukraine, by Andrew Harding

Suzannah Rowntree

June 2025

About the author

Suzannah Rowntree lives in a big house in rural Australia with her wonderful family, drinking fancy tea and writing historical fantasy fiction that blends real-world history with legend, adventure, and a dash of romance.

You can connect with me on:

https://suzannahrowntree.site

Subscribe to my newsletter:

https://subscribepage.io/srauthor

Also by Suzannah Rowntree

The Miss Sharp's Monsters Series

The Werewolf of Whitechapel

A Study in Sirens

Anarchist on the Orient Express

A Vampire in Bavaria

The Miss Dark's Apparitions Series

Tall & Dark

Dark Clouds

Dark & Stormy

Dark & Dawn
A Stab in the Dark
Dark Secrets

The Watchers of Outremer Series

A Wind from the Wilderness
The Lady of Kingdoms
Children of the Desolate
A Day of Darkness
A Conspiracy of Prophets
The House of Mourning
A Stranger in the Land

The Pendragon's Heir Trilogy

The Door to Camelot
The Quest for Carbonek
The Heir of Logres

The Fairy Tale Retold Series

The Rakshasa's Bride
The Prince of Fishes
The Bells of Paradise
Death Be Not Proud
Ten Thousand Thorns
The City Beyond the Glass

Notes

•••

....
....
....
....
....
....
....
....
....
....
....
....
....
....
....
....
....
....
....
....
....
....
....
....
....
....

....
....
....
....
....
....
....
....
....
....
....
....
....
....
....
....
....
....
....
....
....
....
....
....
....
....
....
....
....
....

....
....
....
....
....
....
....
....
....
....
....
....
....
....
....
....
....
....
....
....
....
....
....
....
....
....
....

....
....
....
....
....
....
....
....
....
....
....
....
....
....
....
....
....
....
....
....
....
....
....
....
....
....

....
....
....
....
....
....
....
....
....
....
....
....
....
....
....
....
....
....
....
....
....
....
....
....
....
....

www.ingramcontent.com/pod-product-compliance
Lightning Source LLC
Chambersburg PA
CBHW062145020426
42334CB00020B/2525